Concussions

Fact vs Fiction

WRITTEN BY

Nicholas Staropoli
Research Associate, American Council on Science and Health

A publication of the

AMERICAN COUNCIL
ON SCIENCE AND HEALTH

American Council on Science and Health
1995 Broadway, Suite 202
New York, New York 10023-5860
Tel. (212) 362-7044 • Fax (212) 362-4919
URL: http://www.acsh.org • Email: acsh@acsh.org

Publisher name: American Council on Science and Health
Title: Concussions: Fact Versus Fiction
Author: Nicholas Staropoli
Subject (general): Science and Health
Publication Year: 2016
Binding Type (i.e. perfect (soft) or hardcover): Perfect
ISBN: 978-0-9910055-4-3

Cover image credit to Shutterstock.

Acknowledgements

The American Council on Science and Health appreciates the contributions of the reviewers named below:

Nigel Bark M.D.
Albert Einstein College of Medicine

Jonathan E. Howard M.D.
New York University Langone Medical Center

Hank Campbell
American Council on Science and Health

Josh Bloom Ph.D.
American Council on Science and Health

Table of Contents

Jeb Putzier grew up in Eagle, Idaho, where he attended the Eagle High School and lettered in both basketball and football. After graduating high school, he attended nearby Boise State, where he was able to walk onto the football team as a wide receiver. As a senior he converted[1] to tight end and set several school receiving records at the position. He was selected on the first team in the All Western Athletic Conference and led the nation in touchdowns at that position. After graduation, he was selected in the sixth round of the National Football League (NFL) draft by the Denver Broncos.

Anyone who makes it into the NFL is elite, but even most avid fans probably won't remember much about his career. Though he earned a 5-year $12.5 million from the Colorado franchise he didn't catch a tremendous number of passes and found the end zone only sporadically during seven years in professional football. After being cut by Denver he played in Houston and Seattle, but he never made a pro-bowl and never led the league in any major category. However, there is one statistic in which Jeb Putzier really does stand out: he sustained over one thousand concussions over his entire football career[2].

'Concussion' is a word that is on the mind of everyone in sports in 2015, and with good reason: It is an injury to our most vital organ, the brain. We as a society have become so concerned about injuries that 40 percent of Americans[3] won't let their children play football, a sentiment echoed

---

1   Taubes, G. (1998). The (Political) Science of Salt. Retrieved from https://www.stat.berkeley.edu/~rice/Stat2/salt.html

2   Klis, M. (2015, July 6). NFL Aftermath: Life a medical struggle for Jeb Putzier. Retrieved from http://www.9news.com/story/sports/2015/07/05/jeb-putzier-nfl-brain-injuries/29733075/

3   O'Connor, P. (2014, January 31). Poll Finds 40% Would Sway Children Away From Football. Retrieved from http://www.wsj.com/news/articles/SB100014240527023037436045793531411113118468?mod=rss_US_News

by President Obama[4]. But it's not just football where concussions are an issue; most sports, at all levels, are studying how to reduce them.

Despite the growing attention, there is much yet to be learned about concussion. Due to exaggerated claims (a company called CereScan says they detected "invisible" injuries in Putzier, and for there to be 1,000 concussions in seven years, he has to have gotten one almost every day of the season, including during training camps), what the public believes about concussions is often far different than medical and scientific thinking. In addition, some celebrities and sports stars have also contributed to the confusion of the public by pushing products with unsubstantiated benefits. Thus, we are starting to see the sales of devices such as impact sensors, specialized helmets, and special healing chambers. Some of these may be useful, but many will either be useless, dangerous, or both. Clearly, marketers are trying to profit from raised awareness of concussions.

Some groups are advising caution regardless of the unclear state of some of the science; in November of 2015[5], U.S. Soccer, the governing body for the sport played by more than three million kids in the United States, stated there should be no heading for any players age 10 and under.

How much caution is warranted? That's a puzzle almost as complex as concussions themselves.

4   Remnick, D. (2014, January 27). Going the Distance. Retrieved from http://www.newyorker.com/magazine/2014/01/27/going-the-distance-david-remnick

5   Botelho, G. (2015, November 10). U.S. youth soccer players told: Don't head the ball. Retrieved from http://www.cnn.com/2015/11/10/health/us-youth-soccer-concussions/

1
What is a Concussion?

A traumatic brain injury (TBI) is characterized by a violent movement of the brain inside the cranium, which results in discernable changes to the function of the brain without detectable changes to its structure. Concussions are generally thought of as mild TBIs[6] because they are not life threatening, though as we have seen in athletes and others, they can be severely life altering.

Julie Gilchrist, M.D., of the United States Centers for Disease Control and Prevention, defines a concussion as[7] "a change in the way the brain functions", which can mean almost anything. To-date, there is no known or detectable physical neurological abnormality in the brain that is specific for a concussion.

Concussions cannot be diagnosed with any sort of scan (CT, MRI, X-Ray, etc.) but the notion that they can persists among both the general public, and physicians as well. The American Academy of Neurology guidelines stated in 2013: "CT imaging should not be used to diagnose [sport-related

6   Traumatic Brain Injury (TBI). (2012, June 12). Retrieved from http://www.pattan.net/category/Educational Initiatives/Traumatic Brain Injury (TBI)/page/Traumatic_Brain_Injury_vs_Concussion.htm

7   What is a concussion? (2015, February 16). Retrieved September 15, 2015, from http://www.cdc.gov/headsup/basics/concussion_whatis.html

concussions].” Nonetheless, one study found[8] that athletes who were brought to the hospital with a suspected concussion received a CT scan 75 percent of the time when examined by an emergency medicine physician and 72 percent of the time when examined by a neurologist. However, it is often necessary to preform a CT scan if the physician suspects or needs to rule out other pathologies.

The lack of a definitive diagnostic technique leaves trainers, coaches, and physicians who are trying to evaluate the severity of a head injury, unable to definitively determine whether a concussion has occurred or not. Worse still, concussion symptoms are variable, poorly understood, and often nonspecific.

One of the universal beliefs about concussions is that they occur from a direct blow to the head, however, concussions may occur in the absence of this type of impact. Rather, any impact to the body that causes the head and brain to move rapidly can cause a concussion, for example, whiplash. Another common misconception is that a concussion must involve a loss of consciousness and/or memory. However, according to the CDC and the American Academy of Orthopedic Surgeons[9], less than 10 percent of concussion victims experience a loss of consciousness, and only a third of victims exhibit memory loss.

The symptoms associated[10] with having a concussion form a long and diverse list. A concussed person may exhibit any or none of these

8   Meehan, W. P., d' Hemecourt, P., Collins, C. L., & Comstock, R. D. (2011). Assessment and Management of Sport-Related Concussions in United States High Schools. *The American Journal of Sports Medicine*, 39(11), 2304–2310. http://doi.org/10.1177/0363546511423503

9   Daugherty, K. (2014). Concussions: Doing the Right Thing. Retrieved from http://www.aaos.org/AAOSNow/2014/Jan/clinical/clinical12/?ssopc=1

10 Concussion Signs and Symptoms. (2015, February 16). Retrieved from http://www.cdc.gov/headsup/basics/concussion_symptoms.html

symptoms: inability to recall events prior to or after the impact, mood behavior or personality changes; clumsiness and disorientation; problems concentrating; nausea and vomiting; headaches; and light sensitivity. The length of this list and the nonspecific nature of the symptoms, compound the difficulty of diagnosing the condition. Even more confusing is that concussed persons can often still walk, talk, and interact normally, partly because symptoms of a concussion do not always appear immediately.

The public and many in the sports and medical communities look at concussions the same way they look at infectious diseases or cystic fibrosis, but those conditions all those have a clear cut cause and diagnosis. Concussions do not fit into this type of diagnostic box. Neurologist Dr. Uzma Samadani[11] says there are as many as 43 different definitions of a concussion.

Since no single test can definitively determine if a player has or has not had a concussion, a concussion diagnosis[12] is generally made by a physician using multiple criteria, including symptoms, a neurological examination (balance and coordination) and observations. It's difficult to make an informed decision about a condition when there are so many gaps in our knowledge, but the only safe choice is 'don't play sports', which is hardly a quality recommendation when there are concerns that children are involved in too few activities.

This document will address those concerns.

11 Lapook, J. (2015, January 29). Wile E. Coyote inspires new way to diagnose concussions. Retrieved from http://www.cbsnews.com/news/wile-e-coyote-inspires-new-way-to-diagnose-concussions/

12 Mayo Clinic Staff. (2014, April 2). Diseases and Conditions Concussion Tests and diagnosis. Retrieved from http://www.mayoclinic.org/diseases-conditions/concussion/basics/tests-diagnosis/con-20019272

2

Are concussions on the rise?

There is a near ubiquitous belief that concussions have been on the rise in all contact sports for about a decade, both at the professional and youth levels. There is also a perception that all levels of sports have become more violent. This belief has many, particularly parents of youth sports athletes, worried about the risks of their children participating in sports. However, it is unclear whether concussions are actually increasing or if the increased attention they are receiving is causing an increase in diagnoses.

Due to better diagnosis or more events, the literature is flush with studies that report an increase in concussions in a variety of sports. An estimated 62,000 concussions[13] are sustained each year by college and high school athletes each year in the United States. The Head Health Management System[14] reports that since 2002, concussions have doubled in youth sports. They also note that girls' soccer, boy's hockey, and lacrosse saw the largest increases. A study from the CDC reports that

13 Patient Information. (n.d.). Retrieved from http://www.aans.org/patient information/ conditions and treatments/concussion.aspx

14 Sports Concussion Statistics. (n.d.). Retrieved from http://www.headcasecompany.com/ concussion_info/stats_on_concussions_sports

emergency room visits by children and adolescents for concussion-like symptoms rose 62% between 2001 and 2009[15].

A study in the American Journal of Sports Medicine of youth sports in Virginia over the 11-year period from 1997-2008, found that concussions had risen four fold for the 12 sports studied[16]. A study of 4,000 athletes from 100 different schools, also published in American Journal of Sports Medicine, concluded that from the 2005-06 school year to the 2011-12 school year concussions doubled. According to data from the England Professional Rugby Injury Surveillance Project[17] concussions in professional Rugby had risen over 50 percent from 2008 to 2014, however during this time period total injuries remained fairly constant.

That total injuries have remained rather constant may mean that concussions are not really increasing. Anecdotally, we as a society are more aware of concussions, which have led to coaches and parents scrutinizing each physical contact. Many coaches are now required to undergo training on how to handle athletes with concussions, a practice uncommon a short time ago. According to the National Athletic Trainers' Association, secondary schools in the United States employing an athletic trainer (AT) are also on the rise. In 2005, just 40 percent of secondary

15 Centers for Disease Control and Prevention. (2011). Nonfatal traumatic brain injuries related to sports and recreation activities among persons aged ≤19 years--United States, 2001-2009. Morbidity and Mortality Weekly Report. Retrieved from http://www.ncbi.nlm.nih.gov/pubmed/?term=Nonfatal Traumatic Brain Injuries Related to Sports and Recreation Activities Among Persons Aged ≤19 Years

16 Lincoln AE, Caswell SV, Almquist JL, Dunn RE, Norris JB, Hinton RY. Trends in concussion incidence in high school sports: a prospective 11-year study. Am J Sports Med. 2011 May;39(5):958-63. doi: 10.1177/0363546510392326. Epub 2011 Jan 29. PubMed PMID: 21278427.

17 England Professional Rugby Injury Surveillance Project. (2015, February 1). Retrieved from http://www.englandrugby.com/mm/Document/General/General/01/30/80/08/EnglandProfessionalRugbyInjurySurveillanceProjectReport2013_2014_Neutral.pdf

school employed an AT, but today more than two thirds employ one[18]. All this extra attention on the athletes certainly inflates these rates by catching instances that were missed in the past.

The definition of a concussion has also expanded, and many more symptoms are now considered when deciding if an athlete is concussed. If you go back a few decades, a player would only be considered concussed if they were unconscious, otherwise it was just a "ding[19]." However, as mentioned above, we now know that only about 10 percent of concussions render a person unconscious. The inclusion of other symptoms such as mood and personality changes, light sensitivity, and memory loss, have all contributed to the diagnosis of more concussions.

The increased attention society has given concussions, as well as the expanding definition of the condition, are undoubtedly positive changes. However this attention brings potential pitfalls as well. It has created the perception that sports are more dangerous than they once were, which has led some parents to stop enrolling their children in youth sports at a time when concerns about obesity and physical activity are also on the rise. A 2013 study found that pop-warner football enrollment was down 10 percent from 2010-2012 and according to the Sports & Fitness Industry Association, participation in team sports among children aged 6 through 12 fell from 44.5 percent in 2008 to 40 percent in 2013[20].

---

18 Athletic Trainers Fill a Necessary Niche in Secondary Schools. (2009, March 12). Retrieved from http://www.nata.org/NR031209

19 Bleizeffer, K. (2014, December 23). The changing definition of concussion and why it matters to young athletes. Retrieved from http://wyomingmedicalcenter.org/pulse/2014/12/23/the-changing-definition-of-concussion-and-why-it-matters-to-young-athletes/

20 Fainaru-Wada, M., & Fainaru, S. (2013, November 14). Youth football participation drops. Retrieved from http://espn.go.com/espn/otl/story/_/page/popwarner/pop-warner-youth-football-participation-drops-nfl-concussion-crisis-seen-causal-factor

Youth sports and other organized team activities offer a wide variety of benefits to a child's health and development, so parents should not look at the data that claim concussions are on the rise and assume this means that sports are too dangerous for their child. Instead parents need to realize that sports have always had risks, the same as they did when parents were children. The difference now is that parents armed with the necessary information so they can take the necessary precautions to minimize these risks from impacting their children long term.

Next we need to examine how player impacts are sent for medical follow-up.

3
Sideline tests

The sideline test is one the most important aspects of concussion protocol. A good sideline test can help determine if a player is concussed or if they are safe to return to the field. A bad one will mean immediate referral and better chances for no long-term issues. Fortunately, we have long evolved from the days of "how many fingers am I holding up?" to determine if a player has experienced a concussion. Due to the increased attention the condition has received, all sports leagues now use multi-faceted sideline tests to help identify players who may have sustained a concussion.

The NFL uses a test derived from the Standardized Concussion Assessment Test 2[21] (SCAT2) to asses injured players. The core parameters of this test are memory and concentration. For example, players are given a series of words and numbers, and asked to repeat them. They are also tested for awareness through the use of questions such as, "What month is it?" and "What time is it right now?" Players are scored on these sections and their score is combined with other metrics, including a physical examination, to determine if the player can return to play.

The SCAT2 test not without critics, though. Lawrence Jackson, a linebacker for the Detroit Lions, described earlier this year[22] how he was able to pass this sideline test only to later have physicians diagnose him with a concussion. Jackson told ESPN, "The test was pretty easy if you don't have a significant concussion. It still is. They asked me what day it was, approximately what time it was, what quarter we were in. They wanted me to say the months backwards and a few other things. I got those right. I didn't look disoriented, so they really didn't have any choice but to let me go back in the game."

There is an even more compelling need for better diagnostic tests for concussions. If an athlete sustains a second concussion before he or she has fully healed from the first, they can develop a condition called second impact syndrome (SIS), which causes rapid swelling of the brain and is fatal about 50 percent of the time. A bit more on this would be useful since it is so dramatic. How common is this really? How often do people die?

Another notable flaw in the test is that the score required to determine whether an athlete concussed is uncertain. Data show a high degree of

21 NFL Sideline Concussion Assessment Tool: BASELINE TEST. (n.d.). Retrieved from http://static.nfl.com/static/content/public/photo/2014/02/20/0ap2000000327057.pdf

22 Seifert, K. (2015, February 19). Inside Slant: The plain truth of NFL sideline concussion tests. Retrieved from http://espn.go.com/blog/nflnation/post/_/id/160927/inside-slant-the-plain-truth-of-nfl-sideline-concussion-tests

variability of these scores among non-concussed athletes, particularly on the concentration aspects of the test. These low scores from healthy athletes devalue the low tests scores by possibly concussed athletes. To get around this, athletes are often tested at the start of the season, called their baseline score, and if their post-impact score is lower than baseline it is assumed to be a sign of a concussion. This may sound intuitively correct but the science is not adding up.

Falling below baseline scores as a measure of a concussion was recently studied in popular diagnostic test called ImPACT–the test is considered by experts to be merely a fancier version than the SCAT2. A meta-analysis of the efficacy of the ImPACT found that the test incorrectly diagnosed a concussion in an athlete 22-46 percent of the time[23].

Some critics of baseline analysis also claim that athletes are not sufficiently trustworthy during their baseline examination. There have been reports of athletes attempting to artificially decrease their baseline score. Other critics say that a long period of time between baseline and the post-impact administration, the score may not be reliable; this might be particularly for young people who can experience dramatic shifts in cognitive function from season start to finish.

Clinically and in practice, the King-Devick test (KD)[24] has been shown to be a better diagnostic measure than other sideline tests. The KD test also analyzes other functions of the body known to be affected by concussions: attention, concentration and rapid eye movement The KD test consists of an athlete reading a series of numbers on a card that are unequally spaced on several different lines. The athlete reads the

23 Alsalaheen B, Stockdale K, Pechumer D, Broglio SP. Measurement Error in the Immediate Postconcussion Assessment and Cognitive Testing (ImPACT): Systematic Review. J Head Trauma Rehabil. 2015 Aug 19. [Epub ahead of print] PubMed PMID:26291631.

24 Mayo Clinic. (2014, May 12). King-Devick Test Detects Concussions in Youth Athletes. Retrieved from https://www.youtube.com/watch?v=4_CKo6l9Hss

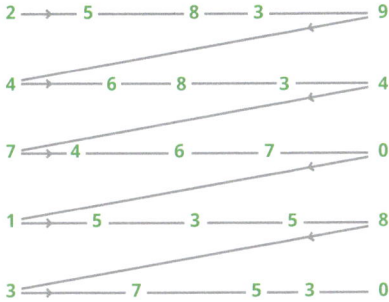

Demonstration Card

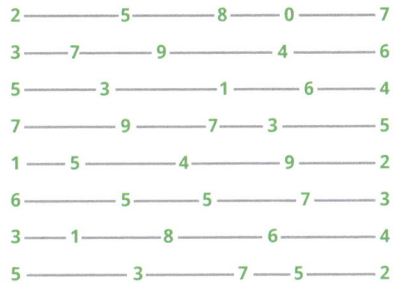

Test 1

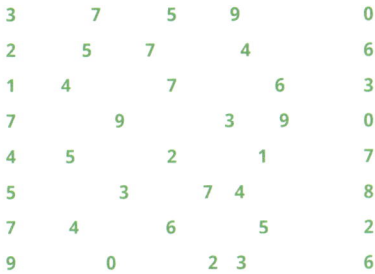

Test 2

Test 3

numbers left to right and the time it takes to do so is measured. This is done three times, and each time the card is different, more difficult, and requires more concentration from the player. A prolonged time to complete the test, as compared to standard times, is indicative of a player who may have a concussion. The increased difficulty of the KD test is likely to prevent players like Lawrence Jackson from slipping through the cracks after a collision, as he did with SCAT2.

For example, a Mayo Clinic study that followed[25] an Arizona youth hockey league for a full season was found to be 100 percent accurate in detecting concussions among the athletes. Another study of youth football[26] players yielded the same results. In a study of professional hockey players, the KD test was able to diagnose concussions in a small sample[27] of athletes that the SCAT2 indicated had no concussion.

Although the KD test also relies on baseline testing to determine if an athlete has sustained a concussion, it is harder for an athlete to fake the baseline score. The test is not familiar nor does it rely on memory. The scientific evidence to-date supports this test's superiority to other tests and more teams should adapt this test.

25 Leong, D. F., Balcer, L. J., Galetta, S. L., Evans, G., Gimre, M., & Watt, D. (2015). The King–Devick test for sideline concussion screening in collegiate football. Journal of Optometry, 8(2), 131–139. http://doi.org/10.1016/j.optom.2014.12.005

26 Seidman DH, Burlingame J, Yousif LR, Donahue XP, Krier J, Rayes LJ, Young R,Lilla M, Mazurek R, Hittle K, McCloskey C, Misra S, Shaw MK. Evaluation of the King-Devick test as a concussion screening tool in high school football players. J Neurol Sci. 2015 Sep15;356(1-2):97-101. doi: 10.1016/j.jns.2015.06.021. Epub 2015 Jun 12. PubMed PMID: 26094155.

27 Galetta MS, Galetta KM, McCrossin J, Wilson JA, Moster S, Galetta SL, Balcer LJ, Dorshimer GW, Master CL. Saccades and memory: baseline associations of the King-Devick and SCAT2 SAC tests in professional ice hockey players. J Neurol Sci. 2013 May 15;328(1-2):28-31. doi: 10.1016/j.jns.2013.02.008. Epub 2013 Mar 15.PubMed PMID: 23499425.

4

Helmets and Sensors

That concussions appear on brain scans is a popular misconception, but an even more prevalent misconception is that helmets can prevent them. There is no scientific evidence that helmets minimize the potential for a concussion. Helmets do prevent life threatening injuries like skull fractures.

Yet one of the primary initiatives of the NFL regarding concussions has been the development of better helmets that minimize the impact of hits to the head. The league has invested millions of dollars in new helmet development, and recently, along with the NFL Player's Association, has funded a study to test and rank new helmets based on safety data of helmets that are used today. However, even the best of these helmets only re-distributes the force of an impact evenly on the skull, it does not stop a concussion.

The holy grail of concussion detection, in recent years, has been the biomedical sensor, which can detect whether an athlete has experienced an impact forceful enough to cause a concussion. Sensors have been long sought after because it is otherwise all but impossible to keep track of every hit a player takes, whereas a sensor in every player's helmet or mouth guard could alert officials when a player has sustained a hit greater than a specified force. Both the U.S. military and professional sports leagues have spent millions of dollars funding grants to encourage research in this area.

Despite their promise, the use of sensors remains very controversial and the sensors themselves have received a high degree of criticism. In

February of this year the NFL[28] suspended a program to monitor impact on players via sensors for many reasons. One being that some league advisers don't believe the sensors can accurately gauge the force of each impact. Another issue is that the data are slow to accumulate, which creates a delay in alerting trainers. It also remains unclear exactly what force is considered dangerous. And force is not the only metric that is a concern. A recent study pointed out[29] that the no sensors that currently on the market could accurately assess angular hits, which are thought to be the most dangerous type. The study also notes that sports in which no helmets are used, such as soccer, where concussions are also very common, helmet sensors would e of no use.

There are many league officials who support the sensor program, but it has been revealed that some of them have conflicts of interest due to being stakeholders in companies that make the sensors. Many colleges use sensors, but their real utility is far from established.

28 Fainaru, S. (2015, February 20). No helmet sensors for NFL in '15. Retrieved from http://espn.go.com/espn/otl/story/_/id/12348395/nfl-teams-use-concussion-sensors-helmets-2015

29 Derek Nevins, Lloyd Smith, Jeff Kensrud, Laboratory Evaluation of Wireless Head Impact Sensor, Procedia Engineering, Volume 112, 2015, Pages 175-179, ISSN 1877-7058, http://dx.doi.org/10.1016/j.proeng.2015.07.195.

5

Can concussions be prevented by rule changes, or better coaching, training and education?

Another strategy to limit concussions involves rule changes, particularly at the youth level. One of these changes that has been discussed is eliminating the practice heading of soccer balls in the youth game. This change has considerable support from USA Women's soccer heroes Brandi Chastain and Cindy Parlow[30], which has led to the change in guidelines for youth players mentioned in the opening. Experts also add this would be particularly beneficial for girls because a girl's neck muscles develop more slowly, making their heads less able to withstand a head injury from whiplash until they are more developed.

The idea of eliminating one aspect of the game is not revolutionary. In youth hockey leagues around the world, body checking is not allowed until pre-teen years at the earliest and Little League baseball has debated

---

30 NBC Bay Area Staff, & Wire Services. (2015, June 15). Brandi Chastain Jumps on Effort to Curb Headers in Youth Soccer. Retrieved from http://www.nbcbayarea.com/news/local/Brandi-Chastain-Jumps-on-Effort-to-Curb-Headers-in-Youth-Soccer-307533571.html

banning curveballs for decades However, some data[31] suggest that when comparing hockey leagues that do and don't enforce this moratorium on checks, there is no difference in the number of head injuries, and light tackling by youth football players until they were much larger and faster might mean more injuries due to inexperience than they would have gotten in youth. For those reasons, eliminating heading in soccer may not lead to decreases in head. Researchers[32] looked at head injuries in both sexes soccer and found that a only a few came from heading the ball; the overwhelming majority have come from player-player contact. Most soccer injuries are things like ankles and knees, not concussions. Still, concussions will certainly not go up with heading removed from the arsenal of techniques vulnerable brains use in the sport.

Though data are unclear and sometimes conflicting, steps are still being taken. The NFL, for example, has implemented several new rules, such as penalizing dangerous hits that target a player's head or defenseless players and revamping the dangerous kick-off play. The results have been encouraging. The number of concussions that have been sustained in regular season games is down 35 percent since 2013, while dangerous hits on defenseless receivers are down 68 percent over the same period! Further investigations are needed to determine if rule changes in soccer and hockey are an effective way to minimize concussions.

At the youth level, USA Football and the NFL have combined to create the Heads Up Football (HUF) initiative and the program has demonstrated

31 Darling SR, Schaubel DE, Baker JG, Leddy JJ, Bisson LJ, Willer B. Intentional versus unintentional contact as a mechanism of injury in youth ice hockey. Br J Sports Med. 2011 May;45(6):492-7. doi: 10.1136/bjsm.2009.063693. Epub 2010 May 19. PubMed PMID: 20484317.

32 Comstock RD, Currie DW, Pierpoint LA, Grubenhoff JA, Fields SK. An Evidence-Based Discussion of Heading the Ball and Concussions in High School Soccer. JAMA Pediatr. 2015 Sep;169(9):830-7. doi:10.1001/jamapediatrics.2015.1062. PubMed PMID: 26168306.

considerable success in reducing concussions in young players. HUF is an educational program that teaches coaches a variety of ways to reduce concussions among their players. These include how to properly fit a player's equipment, how to detect the signs of a concussion, and how to teach proper blocking and tackling techniques to better ensure player safety. Since the program's inception one million players from over 6,000 organizations have participated in HUF. Compared to leagues without HUF training, players with coaches that are HUF certified have 76 percent fewer injuries, 34 percent fewer concussions in practices and 29 percent fewer concussions in games.

These changes are dramatic and so there are calls for youth football leagues around the country to adopt HUF policies and have their coaches certified by HUF. But whether or not those kind of changes will migrate to other sports is unclear. The subject should certainly be examined and evaluated scientifically.

6
Treatment options – real or pseudoscience?

Even the best preventive measures will only serve to reduce the incidence of concussion. They will not eliminate them. The sports community leads in treatment for injuries because sports are lucrative for the players and owners and enjoyed by participants. But when it comes to concussions, treatment advances have not kept pace with diagnostics and prevention. At this time, the only real treatment[33] for concussions is rest and management of symptoms, such as the use of acetaminophen for headaches, a very 19th century approach to a 21st century problem. For significant concussions, a physician may recommend a reduced workload or school load, and concussion sufferers are advised to avoid intense concentration or exertion, but this is far from revolutionary and serves to further highlight the unmet medical need for concussion treatment.

There are several promising ideas that are being investigated. A preliminary study[34] of the use of oral antioxidants showed promise in treating concussion symptoms. Another group at Penn State[35] is working on a cooling helmet based on data that increased brain temperature renders the brain vulnerable to additional injury.

Both are in the very early stages and as is often the case, pseudoscience treatments and outright quackery often fill the void. For example, Seattle Seahawks quarterback Russell Wilson recently tweeted[36] that Recovery Water cured his concussion through their "nanobubble technology". While Wilson may genuinely believe that, it has to be noted that he

33 Mayo Clinic Staff. (2014, April 2). Diseases and Conditions Concussion Treatments and Drugs. Retrieved from http://www.mayoclinic.org/diseases-conditions/concussion/basics/treatment/con-20019272

34 Federation of American Societies for Experimental Biology (FASEB). (2015, April 1). Antioxidant therapy may have promising potential in concussion treatment. ScienceDaily. Retrieved January 17, 2016 from www.sciencedaily.com/releases/2015/04/150401132752.htm

35 Miller, M. (2015, March 5). 'Tip of the iceberg' in concussion treatment. Retrieved from http://news.psu.edu/story/347428/2015/03/05/research/tip-iceberg-concussion-treatment

36 Petchesky, B. (2015, August 26). Aw, Christ. Retrieved from http://deadspin.com/aw-christ-1726772039

is an investor in the company[37]. He does concede that his claim is not backed by "real medical proof." New England Patriots quarterback Tom Brady also once backed a similar drink.

Hyperbaric oxygen therapy is probably the most dangerous pseudo-science treatment for concussions because it has both pseudo-medical and celebrity proponents. This procedure involves a patient entering a chamber where he or she inhales 100 percent oxygen, which is claimed to enhance the body's natural healing powers. Hyperbaric oxygen therapy is approved by the FDA to treat 13 conditions, but not traumatic brain injuries. The Federal Government funded three studies costing taxpayers $70 million[38], all of which concluded these chambers do not increase recovery time or quality of recovery as compared to those who were treated through rest. What's worse is that currently a round of treatment involves 40 visits and can cost a patient anywhere from $5,000 to $12,000[39].

37 Carson, D. (2015, August 26). Russell Wilson Says Recovery Water Healed His Head Injury from NFC Title Game. Retrieved from http://bleacherreport.com/articles/2555730-russell-wilson-says-miracle-water-helped-him-recover-from-a-head-injury

38 Meier, B., & Ivory, D. (2013, July 3). Effective Concussion Treatment Remains Frustratingly Elusive, Despite a Booming Industry. Retrieved from http://www.nytimes.com/2015/07/05/business/effective-concussion-treatment-remains-frustratingly-elu-sive-despite-a-booming-industry.htmlhttp://www.nytimes.com/2015/07/05/business/effective-concussion-treatment-remains-frustratingly-elusive-despite-a-boomin

39 Kime, P. (2014, November 17). Oxygen therapy no better than placebo for treating concussion, study finds. Retrieved from http://www.militarytimes.com/story/military/benefits/health-care/2014/11/17/hyperbaric-oxygen-therapy-concussion-tbi/19183763/

This has not deterred proponents of the treatment. Former New York Jets quarterback Joe Namath[40] is very outspoken about his belief that the treatment is curative. He underwent 120 sessions and claimed that it cured him of the concussions he sustained during his career. Namath has also spent $10 million of his own money to prove the technique has benefits and a clinic in Florida now bears his name in recognition of the support he's given to the treatment regimen. Other proponents, including some physicians, are also intensely lobbying congress to fund more studies.

The situation with concussion treatments is similar to the supplement industry, where the government does not require proof of efficacy provided they use clever language to avoid appearing to be medical. Currently, no science exists that suggests special drinks or chambers work any better than simple rest so none of these products should be allowed to suggest otherwise.

40 Inabinett, M. (2015, July 5). As Ken Stabler's family donates his brain to research, Joe Namath thinks he has an answer for traumatized athletes. Retrieved from http://www.al.com/sports/index.ssf/2015/07/as_ken_stablers_family_leaves.html

7

How we look at concussions must change

Though the landscape for prevention and diagnosis of concussions has exploded during the past five years, there remains a long way to go. Though results are inconclusive and/or questionable today, there remains hope that investments in basic research and technology will lead to revolutions in diagnostics, prevention and even treatments.

We also need to change the way we see concussions as a society. For too long, American culture has mocked or downplayed concussions and head injuries in general. This attitude appears in several forms, such as the NFL profiting off videos[41] of "skull-crushing hits" and Denver Broncos quarterback Peyton Manning[42] on Saturday Night Live joking about being "hit in the head a lot." When we as a society marginalize these injuries it manifests in dangerous ways. For example, a study[43] recently found that 75 percent of surveyed rugby players said they would continue playing in

---

41 Special Order Catalogue: Crunch and Hard Hitting. (n.d.). Retrieved from http://www.nflfilms.com/specialorders/crunchHardHitting.php

42 Strachan, M. (2015, February 16). Peyton Manning Made An Unfunny Joke About Getting 'Hit In The Head A Lot' Last Night. Retrieved from http://www.huffingtonpost.com/2015/02/16/peyton-manning-joke-hit-in-the-head_n_6693672.html

43 O'Connell E, Molloy MG. Concussion in rugby: knowledge and attitudes of players. Ir J Med Sci. 2015 May 31. [Epub ahead of print] PubMed PMID: 26026952.

a game even if they knew they had a concussion. Another found[44] that a third of high school football players felt they had sustained a concussion but did not seek medical help.

Concussions in our society are still not regarded with the same concern as other neurological disorders, and this leads our athletes, particularly the young ones, to not take them seriously. In effect, society has fostered the mentality that concussions are an inevitable part of sports. If we want to mitigate this serious but often-overlooked injury, we don't just need to advance our technology, we need to advance the way we think and talk about it as well.

44 Israel, M. (2012). Awareness and Attitudes of High School Athletes towards Concussions. Retrieved from https://www.researchgate.net/publication/267912786_Awareness_and_Attitudes_of_High_School_Athletes_towards_Concussions

Conclusion

Throughout the last decade, concussions have received a tremendous amount of attention by the media, sports leagues, the medical community and the parents of young athletes. This has led to much confusion on what a concussion is, how it is diagnosed is and how to prevent them. Because of this new focus, our understanding of the condition has greatly improved, however there is still much to be done. The future may bring new advances in biosensors, helmets and diagnostic tests but in the meantime efforts to reduce the impact of concussions on athletes and society should focus on rule changes, behavior changes, improved coaching and education, incorporating the KD test into sideline protocols and governmental crackdown on treatments that lack legitimate science.

Until then, the benefits of sports are enormous; life is not risk free. This paper shows how the risks of concussion can be reduced. Let your kids play but make sure what can be done to prevent concussion is being done and if there is any possibility they might be concussed make sure they stop playing and rest.

References

Alsalaheen B, Stockdale K, Pechumer D, Broglio SP. Measurement Error in the Immediate Postconcussion Assessment and Cognitive Testing (ImPACT): Systematic Review. J Head Trauma Rehabil. 2015 Aug 19. [Epub ahead of print] PubMed PMID:26291631.

Athletic Trainers Fill a Necessary Niche in Secondary Schools. (2009, March 12). Retrieved from http://www.nata.org/NR031209

Bleizeffer, K. (2014, December 23). The changing definition of concussion and why it matters to young athletes. Retrieved from http://wyomingmedicalcenter.org/pulse/2014/12/23/the-changing-definition-of-concussion-and-why-it-matters-to-young-athletes/

Botelho, G. (2015, November 10). U.S. youth soccer players told: Don't head the ball. Retrieved from http://www.cnn.com/2015/11/10/health/us-youth-soccer-concussions/

Carson, D. (2015, August 26). Russell Wilson Says Recovery Water Healed His Head Injury from NFC Title Game. Retrieved from http://bleacherreport.com/articles/2555730-russell-wilson-says-miracle-water-helped-him-recover-from-a-head-injury

Centers for Disease Control and Prevention. (2011). Nonfatal traumatic brain injuries related to sports and recreation activities among persons aged ≤19 years--United States, 2001-2009. Morbidity and Mortality Weekly Report. Retrieved from http://www.ncbi.nlm.nih.gov/pubmed/?term=Nonfatal Traumatic Brain Injuries Related to Sports and Recreation Activities Among Persons Aged ≤19 Years

Concussion Signs and Symptoms. (2015, February 16). Retrieved from http://www.cdc.gov/headsup/basics/concussion_symptoms.html

Comstock RD, Currie DW, Pierpoint LA, Grubenhoff JA, Fields SK. An Evidence-Based Discussion of Heading the Ball and Concussions in High School Soccer. JAMA Pediatr. 2015 Sep;169(9):830-7. doi:10.1001/jamapediatrics.2015.1062. PubMed PMID: 26168306.

Darling SR, Schaubel DE, Baker JG, Leddy JJ, Bisson LJ, Willer B. Intentional versus unintentional contact as a mechanism of injury in youth ice hockey. Br J Sports Med. 2011

May;45(6):492-7. doi: 10.1136/bjsm.2009.063693. Epub 2010 May 19. PubMed PMID: 20484317.

Derek Nevins, Lloyd Smith, Jeff Kensrud, Laboratory Evaluation of Wireless Head Impact Sensor, Procedia Engineering, Volume 112, 2015, Pages 175-179, ISSN 1877-7058, http://dx.doi.org/10.1016/j.proeng.2015.07.195.

Daugherty, K. (2014). Concussions: Doing the Right Thing. Retrieved from http://www.aaos.org/AAOSNow/2014/Jan/clinical/clinical12/?ssopc=1

England Professional Rugby Injury Surveillance Project. (2015, February 1). Retrieved from http://www.englandrugby.com/mm/Document/General/General/01/30/80/08/EnglandProfessionalRugbyInjurySurveillanceProjectReport2013_2014_Neutral.pdf

Fainaru, S. (2015, February 20). No helmet sensors for NFL in '15. Retrieved from http://espn.go.com/espn/otl/story/_/id/12348395/nfl-teams-use-concussion-sensors-helmets-2015

Fainaru-Wada, M., & Fainaru, S. (2013, November 14). Youth football participation drops. Retrieved from http://espn.go.com/espn/otl/story/_/page/popwarner/pop-warner-youth-football-participation-drops-nfl-concussion-crisis-seen-causal-factor

Federation of American Societies for Experimental Biology (FASEB). (2015, April 1). Antioxidant therapy may have promising potential in concussion treatment. ScienceDaily. Retrieved January 17, 2016 from www.sciencedaily.com/releases/2015/04/150401132752.htm

Galetta MS, Galetta KM, McCrossin J, Wilson JA, Moster S, Galetta SL, Balcer LJ, Dorshimer GW, Master CL. Saccades and memory: baseline associations of the King-Devick and SCAT2 SAC tests in professional ice hockey players. J Neurol Sci. 2013 May 15;328(1-2):28-31. doi: 10.1016/j.jns.2013.02.008. Epub 2013 Mar 15.PubMed PMID: 23499425.

Inabinett, M. (2015, July 5). As Ken Stabler's family donates his brain to research, Joe Namath thinks he has an answer for traumatized athletes. Retrieved from http://www.al.com/sports/index.ssf/2015/07/as_ken_stablers_family_leaves.html

Israel, M. (2012). Awareness and Attitudes of High School Athletes towards Concussions. Retrieved from https://www.researchgate.net/publication/267912786_Awareness_and_Attitudes_of_High_School_Athletes_towards_Concussions

Kime, P. (2014, November 17). Oxygen therapy no better than placebo for treating concussion, study finds. Retrieved from http://www.militarytimes.com/story/military/benefits/health-care/2014/11/17/hyperbaric-oxygen-therapy-concussion-tbi/19183763/

Klis, M. (2015, July 6). NFL Aftermath: Life a medical struggle for Jeb Putzier. Retrieved from http://www.9news.com/story/sports/2015/07/05/jeb-putzier-nfl-brain-injuries/29733075/

Lapook, J. (2015, January 29). Wile E. Coyote inspires new way to diagnose concussions. Retrieved from http://www.cbsnews.com/news/wile-e-coyote-inspires-new-way-to-diagnose-concussions/

Leong, D. F., Balcer, L. J., Galetta, S. L., Evans, G., Gimre, M., & Watt, D. (2015). The King–Devick test for sideline concussion screening in collegiate football. Journal of Optometry, 8(2), 131–139. http://doi.org/10.1016/j.optom.2014.12.005

Lincoln AE, Caswell SV, Almquist JL, Dunn RE, Norris JB, Hinton RY. Trends in concussion incidence in high school sports: a prospective 11-year study. Am J Sports Med. 2011 May;39(5):958-63. doi: 10.1177/0363546510392326. Epub 2011 Jan 29. PubMed PMID: 21278427.

Mayo Clinic. (2014, May 12). King-Devick Test Detects Concussions in Youth Athletes. Retrieved from https://www.youtube.com/watch?v=4_CKo6l9Hss

Mayo Clinic Staff. (2014, April 2). Diseases and Conditions Concussion Tests and diagnosis. Retrieved from http://www.mayoclinic.org/diseases-conditions/concussion/basics/tests-diagnosis/con-20019272

Mayo Clinic Staff. (2014, April 2). Diseases and Conditions Concussion Treatments and Drugs. Retrieved from http://www.mayoclinic.org/diseases-conditions/concussion/basics/treatment/con-20019272

Meehan, W. P., d' Hemecourt, P., Collins, C. L., & Comstock, R. D. (2011). Assessment and Management of Sport-Related Concussions in United States High Schools. *The American Journal of Sports Medicine*, 39(11), 2304–2310. http://doi.org/10.1177/0363546511423503

Meier, B., & Ivory, D. (2013, July 3). Effective Concussion Treatment Remains Frustratingly Elusive, Despite a Booming Industry. Retrieved from http://www.nytimes.com/2015/07/05/business/effective-concussion-treatment-remains-frustratingly-elusive-despite-a-booming-industry.html

Miller, M. (2015, March 5). 'Tip of the iceberg' in concussion treatment. Retrieved from http://news.psu.edu/story/347428/2015/03/05/research/tip-iceberg-concussion-treatment

NBC Bay Area Staff, & Wire Services. (2015, June 15). Brandi Chastain Jumps on Effort to Curb Headers in Youth Soccer. Retrieved from http://www.nbcbayarea.com/news/local/Brandi-Chastain-Jumps-on-Effort-to-Curb-Headers-in-Youth-Soccer-307533571.html

NFL Sideline Concussion Assessment Tool: BASELINE TEST. (n.d.). Retrieved from http://static.nfl.com/static/content/public/photo/2014/02/20/0ap2000000327057.pdf

O'Connell E, Molloy MG. Concussion in rugby: knowledge and attitudes of players. Ir J Med Sci. 2015 May 31. [Epub ahead of print] PubMed PMID: 26026952.

O'Connor, P. (2014, January 31). Poll Finds 40% Would Sway Children Away From Football. Retrieved from http://www.wsj.com/news/articles/SB10001424052702303743604579353141111318468?mod=rss_US_News

Patient Information. (n.d.). Retrieved from http://www.aans.org/patient information/conditions and treatments/concussion.aspx

Petchesky, B. (2015, August 26). Aw, Christ. Retrieved from http://deadspin.com/aw-christ-1726772039

Remnick, D. (2014, January 27). Going the Distance. Retrieved from http://www.newyorker.com/magazine/2014/01/27/going-the-distance-david-remnick

Seidman DH, Burlingame J, Yousif LR, Donahue XP, Krier J, Rayes LJ, Young R,Lilla M, Mazurek R, Hittle K, McCloskey C, Misra S, Shaw MK. Evaluation of the King-Devick test as a concussion screening tool in high school football players. J Neurol Sci. 2015 Sep15;356(1-2):97-101. doi: 10.1016/j.jns.2015.06.021. Epub 2015 Jun 12. PubMed PMID: 26094155.

Seifert, K. (2015, February 19). Inside Slant: The plain truth of NFL sideline concussion tests. Retrieved from http://espn.go.com/blog/nflnation/post/_/id/160927/inside-slant-the-plain-truth-of-nfl-sideline-concussion-tests

Special Order Catalogue: Crunch and Hard Hitting. (n.d.). Retrieved from http://www.nflfilms.com/specialorders/crunchHardHitting.php

Sports Concussion Statistics. (n.d.). Retrieved from http://www.headcasecompany.com/concussion_info/stats_on_concussions_sports

Strachan, M. (2015, February 16). Peyton Manning Made An Unfunny Joke About Getting 'Hit In The Head A Lot' Last Night. Retrieved from http://www.huffingtonpost.com/2015/02/16/peyton-manning-joke-hit-in-the-head_n_6693672.html

Taubes, G. (1998). The (Political) Science of Salt. Retrieved from https://www.stat.berkeley.edu/~rice/Stat2/salt.html

Traumatic Brain Injury (TBI). (2012, June 12). Retrieved from http://www.pattan.net/category/Educational Initiatives/Traumatic Brain Injury (TBI)/page/Traumatic_Brain_Injury_vs_Concussion.htm

What is a concussion? (2015, February 16). Retrieved September 15, 2015, from http://www.cdc.gov/headsup/basics/concussion_whatis.html

www.ingramcontent.com/pod-product-compliance
Lightning Source LLC
Chambersburg PA
CBHW041718200326
41520CB00001B/148